Weather Update

Storms

by Terri Sievert

Consultant:
Joseph M. Moran, PhD
Associate Director, Education Program
American Meteorological Society, Washington, D.C.

Mankato, Minnesota

Bridgestone Books are published by Capstone Press,
151 Good Counsel Drive, P.O. Box 669, Mankato, Minnesota 56002.
www.capstonepress.com

Library of Congress Cataloging-in-Publication Data
Sievert, Terri.
    Storms / by Terri Sievert.
    p. cm.—(Bridgestone books. Weather update)
    Includes bibliographical references and index.
    ISBN 0-7368-3738-8 (hardcover)
    1. Storms—Juvenile literature. I. Title. II. Series.
QC941.3.S53 2005
551.55—dc22
    2004010846

Summary: Describes storms, including thunderstorms, tropical storms, hurricanes, tornadoes, blizzards, and monsoons.

Editorial Credits
Christopher Harbo, editor; Molly Nei, set designer; Wanda Winch, photo researcher;
    Scott Thoms, photo editor

Photo Credits
Corbis/Annie Griffiths Belt, 10; Jim Reed, 8; Reuters, 14, 18; Tony Arruza, 12
Dan Delaney Photography, cover (child), back cover
Digital Vision/Jim Reed, cover (background), 1, 6
Folio Inc./Charlie Archambault, 20
Leanne W. Richard, 4
Photodisc/StockTrek, 16

1 2 3 4 5 6 10 09 08 07 06 05

Table of Contents

What Is a Storm? 5

Lightning and Thunder 7

Tornadoes 9

Blizzards 11

Tropical Cyclones 13

Tropical Storms 15

Hurricanes 17

Monsoons 19

Storm Damage 21

Glossary 22

Read More 23

Internet Sites 23

Index 24

What Is a Storm?

Lightning flashes. Thunder rumbles. Large raindrops slap against the window. A storm has arrived.

Storms form when warm and cold air meet. Cold air pushes under the warm air. It forces warm air to rise. Wind is created by the moving air. As it rises, the warm air cools. **Water vapor** in the air **condenses**, and clouds form. Raindrops or snowflakes fall.

◀ The sky turns dark gray as a thunderstorm approaches.

Lightning and Thunder

Thunderstorms bring lightning, thunder, and rain. Sometimes thunderstorms bring **hail** or snow.

Lightning forms when **electricity** builds up in a storm cloud. The electricity in the cloud is drawn to electricity in the ground. A **leader** of electricity reaches down from the cloud. The leader meets electricity rising from the ground. A lightning strike is made.

The air around the lightning becomes hot. The heated air creates a **sound wave**. People hear the sound wave as thunder.

◀ A lightning bolt stretches from the clouds to the ground on a stormy night.

Tornadoes

A tornado is a spinning column of air. Tornadoes form along the edges of very strong thunderstorms. Tornadoes form when warm air is quickly swept upward. The air spins into a long column as it rises. The column of air reaches from the ground to the storm clouds.

A tornado's winds can reach 250 miles (402 kilometers) per hour or more. A tornado can pick up animals and uproot trees. It can destroy buildings.

◄ A long thin tornado looks like a rope as it spins across Nebraska farmland.

Blizzards

Blizzards are very strong winter storms. They form when cold air meets warmer air. The warmer air quickly rises. Strong winds blow, and heavy snow falls. The winds blow snow into large drifts that block roads. Schools and businesses sometimes close because of heavy snow during blizzards.

Air temperatures are very cold during a blizzard. Temperatures are usually 20 degrees Fahrenheit (minus 7 degrees Celsius) or lower. Strong winds make the air feel even colder.

◀ A man walks through drifting snow after driving off the road in a North Dakota blizzard.

Tropical Cyclones

Tropical **cyclones** form over the ocean. Most of these storms form near the equator. The equator is an imaginary line around the center of earth.

Tropical cyclones have swirling winds. Storm clouds spin around a central point. Thunderstorms form and heavy rains fall.

A weak tropical cyclone is a tropical **depression**. Its winds are less than 39 miles (63 kilometers) per hour.

◀ Strong winds cause white-capped waves during tropical cyclone Gordon in Palm Beach, Florida in November 1994.

14

Tropical Storms

A tropical storm is a strong tropical cyclone. Tropical storms bring heavy rains over the ocean and land. Wind speeds rise above 39 miles (63 kilometers) per hour. The winds damage buildings. Heavy rains cause flooding.

Tropical storms are named to help people tell them apart. Tropical storms in the Atlantic Ocean get men's or women's names. Tropical storms in the Western Pacific Ocean are named after flowers or animals.

◄ Police block a street as tropical storm Gabrielle's heavy rains and high winds hit southern Florida in September 2001.

Hurricanes

Hurricanes are the strongest tropical storms. Storm clouds twist around a calm center called the eye. Winds outside the eye are very strong. They blow 74 miles (119 kilometers) per hour or faster. Winds inside the eye are much weaker.

Hurricane winds can cause great damage on land. Storm surges are one damaging effect of hurricane winds. A storm surge is water that is pushed toward shore by hurricane winds. It floods land and damages buildings. A storm surge can be 3 to 20 feet (1 to 6 meters) high.

◄ A hurricane looks like a giant pinwheel as it spins toward the southeastern coast of the United States.

Monsoons

Monsoons are winds that blow in southern Asia during spring and summer. As these winds blow from sea to land, they bring heavy rain.

Monsoon rains form when the sun warms the land. Warm air above the land rises. Cool, moist wind blows inland from the sea. Where the warm and cool air meet, rain clouds form. A large amount of rain falls in a short period of time. The rain floods the land.

◄ People wade through flooded streets in India after days of heavy monsoon rains.

Storm Damage

Major storms cause a great deal of damage. They can also be deadly. In 1999, more than 70 tornadoes hit Oklahoma and Kansas in one day. Thousands of homes were damaged. The tornadoes killed 49 people.

In 1992, Hurricane Andrew hit Florida and Louisiana. Andrew's winds rose above 140 miles (225 kilometers) per hour. Damage from the storm cost billions of dollars.

Thunderstorms, tornadoes, and hurricanes can be dangerous. Listen to the news when a storm is near. You can learn how to stay safe.

◄ Cars, trees, and homes were destroyed when tornadoes tore through Oklahoma in 1999.

Glossary

condense (kuhn-DENSS)—to turn from a gas into a liquid

cyclone (SYE-clone)—a storm with strong winds that blow around a quiet center

depression (di-PRESH-uhn)—an area of low air pressure that brings precipitation

electricity (e-lek-TRISS-uh-tee)—a form of energy caused by moving particles

hail (HAYL)—balls of ice that fall during some thunderstorms

leader (LEED-ur)—a stream of electricity that reaches from a cloud to the ground to form lightning

sound wave (SOUND WAYV)—a series of vibrations in air, solids, or liquids that can be heard

water vapor (WAH-tur VAY-pur)—water in gas form; water vapor is one of the many invisible gases in air.

Read More

Maslin, Mark. *Storms.* Restless Planet. Austin, Texas: Raintree Steck-Vaughn, 2000.

O'Hare, Ted. *Storms.* Weather Report. Vero Beach, Fla.: Rourke, 2003.

Internet Sites

FactHound offers a safe, fun way to find Internet sites related to this book. All of the sites on FactHound have been researched by our staff.

Here's how:
1. Visit *www.facthound.com*
2. Type in this special code **0736837388** for age-appropriate sites. Or enter a search word related to this book for a more general search.
3. Click on the **Fetch It** button.

FactHound will fetch the best sites for you!

Index

blizzards, 11

electricity, 7

flooding, 15, 17, 19
formation of storms, 5, 9,
    11, 13, 19

hail, 7
Hurricane Andrew, 21
hurricanes, 17, 21

lightning, 5, 7

monsoons, 19

naming storms, 15

rain, 5, 7, 13, 15, 19

snow, 5, 7, 11
storm damage, 9, 15, 17, 21
storm surges, 17

thunder, 5, 7
thunderstorms, 7, 9, 13, 21
tornadoes, 9, 21
tropical cyclones, 13, 15
tropical depression, 13
tropical storms, 15, 17

water vapor, 5
wind, 5
    storm winds, 9, 11, 13,
        15, 17, 19, 21